Tech Career Mastery Navigating the Path to Professional Growth in the Digital World

Damien Ellis

Copyright © [2023]

Title: Tech Career Mastery Navigating the Path to Professional Growth in the Digital World
Author's: Damien Ellis

This book was printed and published by [Publisher's: **Damien Ellis**] in [2023]

ISBN:

TABLE OF CONTENT

TABLE OF CONTENT

Chapter 5: Thriving in the Tech Workplace 46

Chapter 6: Embracing Leadership and Entrepreneurship in Tech 56

Chapter 1: The Digital Landscape: An Introduction to the Tech Industry

The Evolution of the Tech Industry

In today's fast-paced digital world, the tech industry is constantly evolving and transforming at an unprecedented rate. From the early days of punch cards and mainframe computers to the current era of artificial intelligence and blockchain technology, the tech industry has come a long way. This subchapter explores the fascinating journey of the tech industry's evolution and its implications for career growth.

The tech industry has always been at the forefront of innovation, driving societal progress and transforming the way we live, work, and communicate. Over the years, we have witnessed remarkable advancements in various fields, including software development, hardware engineering, data analysis, cybersecurity, and more. These advancements have not only revolutionized industries but also opened up a plethora of career opportunities for professionals seeking growth and development in the tech sector.

One of the key drivers of the tech industry's evolution has been the exponential growth of computing power. The introduction of microprocessors and personal computers in the 1970s paved the way for a democratization of technology, making it accessible to individuals and businesses alike. This led to the rise of software development and the birth of iconic companies like Microsoft and Apple.

In the 1990s, the advent of the internet brought about another wave of transformation, connecting people across the globe and enabling the creation of countless online platforms and services. This era witnessed the emergence of e-commerce, social media, and search engines, fundamentally altering the way businesses operate and individuals interact. As a result, career opportunities in web development, digital marketing, and data analytics skyrocketed.

More recently, we have witnessed the rapid growth of emerging technologies such as artificial intelligence, machine learning, and blockchain. These technologies have the potential to revolutionize industries like healthcare, finance, transportation, and more. Professionals with expertise in these areas are in high demand, making it an exciting time for career growth in the tech industry.

However, keeping up with the evolution of the tech industry requires continuous learning and adaptation. Employers are looking for individuals who can embrace change, acquire new skills, and stay up to date with the latest trends and advancements. To navigate the path to professional growth in the digital world, it is crucial to seek out opportunities for lifelong learning, attend industry conferences, join professional networks, and invest in professional development courses.

In conclusion, the tech industry has evolved significantly over the years, presenting abundant opportunities for career growth. Professionals who are willing to embrace change, adapt to new technologies, and continuously learn will be well-positioned to thrive in this dynamic industry. By staying informed, networking with like-minded individuals, and investing in their own professional development, employees can navigate the ever-changing landscape of the tech industry and achieve mastery in their tech careers.

Understanding the Digital Transformation

In today's rapidly evolving digital world, the concept of digital transformation has become increasingly important for professionals seeking career growth. The term "digital transformation" refers to the integration of digital technology into all aspects of an organization, fundamentally changing how businesses operate and deliver value to their customers. This subchapter aims to provide a comprehensive understanding of the digital transformation and its implications for career growth.

The digital transformation has revolutionized industries across the globe, disrupting traditional business models and creating new opportunities. Employers now require a new set of skills and competencies to navigate this digital landscape successfully. Understanding the digital transformation is crucial for employees who wish to remain relevant and thrive in their careers.

Firstly, this subchapter will delve into the reasons behind the digital transformation. It will explore how advancements in technology, such as artificial intelligence, big data, and cloud computing, have prompted organizations to adopt digital strategies to stay competitive. Additionally, it will highlight the benefits that digital transformation brings, such as increased efficiency, improved customer experiences, and enhanced decision-making.

Furthermore, this subchapter will discuss the impact of digital transformation on various industries. It will demonstrate how traditional roles are evolving and how new positions are emerging, requiring professionals to upskill or reskill themselves. The chapter will provide insights into the key skills and knowledge areas that are in high demand, including data analytics, digital marketing, cybersecurity, and project management.

Additionally, the subchapter will address the challenges and opportunities that the digital transformation presents to employees. It will discuss the potential risks of job displacement due to automation and explain how individuals can proactively adapt and thrive in the digital era. It will provide guidance on how employees can leverage technology to enhance their productivity, creativity, and overall job satisfaction.

Lastly, this subchapter will provide practical tips and strategies for career growth in the digital world. It will emphasize the importance of continuous learning and professional development to stay ahead in this rapidly changing landscape. It will explore resources, such as online courses, certifications, and networking platforms, that employees can utilize to enhance their knowledge and skills.

In conclusion, understanding the digital transformation is crucial for career growth in the digital world. This subchapter aims to equip employees with the knowledge, skills, and strategies necessary to navigate the digital transformation successfully and thrive in their professional lives. By embracing the digital transformation and embracing opportunities for growth, employees can position themselves for long-term success in the digital era.

The Importance of Tech Careers in Today's World

In today's rapidly evolving digital landscape, the importance of tech careers cannot be overstated. The world is becoming increasingly reliant on technology, and businesses across all industries are looking to harness the power of digital tools and solutions to stay competitive. This shift has created a tremendous demand for skilled tech professionals who can navigate this digital realm and drive innovation.

For employees seeking career growth, embracing a tech career is a strategic move. The tech industry offers a wealth of opportunities for advancement, as well as the chance to work on cutting-edge projects that have a global impact. By entering the tech field, individuals can position themselves at the forefront of technological advancements, ensuring their skills remain relevant in an ever-changing job market.

One of the key reasons why tech careers are so vital today is the increasing reliance on automation and artificial intelligence (AI). As businesses strive to optimize processes and improve efficiency, the need for professionals who can develop and manage these technologies has skyrocketed. Tech careers encompass a wide range of roles, from software developers and data scientists to cybersecurity analysts and cloud architects. The demand for these roles is only expected to grow as businesses continue to invest in digital transformation initiatives.

Tech careers also offer a high degree of flexibility and mobility. With the rise of remote work and the global nature of the industry, tech professionals have the opportunity to work with companies around the world, collaborating with diverse teams and gaining exposure to different cultures and ways of thinking. This global connectivity allows for unparalleled networking and learning opportunities, facilitating career growth and personal development.

Furthermore, tech careers often come with attractive compensation packages. Due to the scarcity of skilled tech professionals, companies are willing to offer competitive salaries and benefits to attract and retain top talent. This financial stability is an appealing aspect for those looking to secure their future and achieve long-term financial success.

In conclusion, the importance of tech careers in today's world cannot be ignored. For employees seeking career growth, the tech industry offers a wealth of opportunities, from cutting-edge projects to global collaboration. With the increasing reliance on technology, skilled tech professionals are in high demand, making it a strategic choice for individuals looking to secure their future in the digital world. By embracing a tech career, individuals can position themselves at the forefront of innovation, ensuring their skills remain valuable and adaptable in an ever-evolving job market.

Chapter 2: Exploring Different Tech Career Paths

Software Development and Engineering

In today's digital world, software development and engineering have become pivotal fields that drive innovation and shape the future of technology. As the demand for cutting-edge software solutions continues to rise, so does the need for skilled professionals who can navigate the complex world of coding, testing, and deployment. In this subchapter, we will explore the world of software development and engineering and provide valuable insights for employees seeking career growth in this dynamic industry.

Software development is the process of designing, coding, testing, and maintaining software systems. It involves a wide range of skills and expertise, including programming languages, algorithms, data structures, and software architecture. Engineers in this field are responsible for creating software applications that meet specific requirements, are scalable, and perform efficiently. They work closely with stakeholders, such as product managers and designers, to understand user needs and translate them into functional software solutions.

To excel in software development and engineering, employees should continuously update their skills and stay up-to-date with the latest technologies and methodologies. The software industry is constantly evolving, and professionals must be adaptable to change. Continuous learning through workshops, online courses, or industry conferences is crucial for staying competitive in this fast-paced environment.

Moreover, software development is a collaborative process, and effective communication and teamwork skills are essential. Working in cross-functional teams, engineers must collaborate with designers, product managers, and quality assurance professionals to ensure the successful delivery of software products. Clear communication, empathy, and the ability to work well with others are highly valued skills in this field.

In addition to technical skills, employees should also possess problem-solving abilities and a strong attention to detail. Software development often involves solving complex problems, and engineers must be able to analyze and break down problems into manageable components. Attention to detail is crucial for writing bug-free code and ensuring that software systems function as intended.

As technology continues to advance, software development and engineering offer numerous opportunities for career growth. Employees can specialize in various areas, such as web development, mobile app development, data engineering, or artificial intelligence. Additionally, leadership roles, such as software architect or engineering manager, provide avenues for advancement and increased responsibility.

In conclusion, software development and engineering are dynamic and exciting fields that offer employees vast opportunities for career growth. By continuously updating their skills, fostering effective communication, and embracing collaboration, individuals can thrive in this ever-evolving industry. Whether it's creating innovative software solutions or leading teams towards success, software development and engineering provide an exciting path for professional growth in the digital world.

Data Science and Analytics

In today's digital world, data has become the new currency. Every industry, from finance to healthcare, relies heavily on data to make informed decisions and gain a competitive edge. As a result, the demand for professionals skilled in data science and analytics has skyrocketed. This subchapter explores the realm of data science and analytics, shedding light on its importance, career opportunities, and strategies for professional growth.

Data science is the interdisciplinary field that uses scientific methods, processes, algorithms, and systems to extract knowledge and insights from structured and unstructured data. It involves a combination of programming, statistics, mathematics, and domain knowledge to analyze and interpret data. Data analytics, on the other hand, focuses on extracting meaningful patterns and insights from data to drive business decisions.

The first section of this subchapter delves into the significance of data science and analytics in today's job market. With the exponential growth of data, organizations are increasingly relying on data-driven strategies to gain a competitive advantage. Employers are seeking professionals who can sift through vast amounts of data, identify trends, and provide actionable insights. By mastering data science and analytics skills, individuals can position themselves for career growth and open up new opportunities in various industries.

The second section explores the diverse career paths available in data science and analytics. From data scientists and analysts to business intelligence specialists and machine learning engineers, there is a wide range of roles to choose from. This subchapter provides an overview of each role, discussing the required skills, responsibilities, and potential growth prospects. Additionally, it highlights the industries that heavily rely on data science and analytics, such as finance, healthcare, e-commerce, and marketing.

Lastly, this subchapter offers practical advice and strategies for professionals looking to advance their careers in data science and analytics. It provides insights on acquiring the necessary skills through online courses, boot camps, and certifications. It also emphasizes the importance of continuous learning and staying up-to-date with the latest tools and technologies in the field. Additionally, it covers networking opportunities, joining professional organizations, and showcasing one's work through projects and portfolios.

In conclusion, data science and analytics have become essential for career growth in the digital world. This subchapter serves as a guide for professionals looking to navigate the path to success in this field. By understanding the importance of data science, exploring various career paths, and adopting strategies for professional growth, individuals can position themselves as valuable assets in the data-driven economy.

Cybersecurity and Information Assurance

In today's rapidly advancing digital world, the importance of cybersecurity and information assurance cannot be overstated. As technology continues to shape and transform various industries, the need for skilled professionals in the field of cybersecurity has become more critical than ever before. This subchapter aims to shed light on the significance of cybersecurity and information assurance for employees, particularly those seeking career growth in the digital landscape.

1. Understanding Cybersecurity: Cybersecurity refers to the practice of protecting computer systems, networks, and data from digital threats. It involves implementing measures to prevent unauthorized access, use, disclosure, disruption, modification, or destruction of information. With the increasing number of cyberattacks and data breaches, organizations across all industries are recognizing the need to prioritize cybersecurity to safeguard their sensitive information and maintain the trust of their customers.

2. The Role of Information Assurance: Information assurance, on the other hand, focuses on ensuring the confidentiality, integrity, and availability of data. It encompasses the processes, policies, and technologies used to protect and manage information against risks such as unauthorized access, data corruption, and system failures. By implementing robust information assurance practices, companies can effectively safeguard their critical information assets and maintain operational continuity.

3. The Growing Demand for Cybersecurity Professionals: As cyber threats become more sophisticated, the demand for

cybersecurity professionals continues to rise. Organizations are actively seeking skilled individuals who can identify vulnerabilities, develop effective security strategies, and respond promptly to security incidents. By developing expertise in cybersecurity and information assurance, employees can enhance their career prospects and unlock exciting opportunities in this high-demand field.

4. Navigating the Path to Professional Growth: To excel in the cybersecurity domain, employees must continuously update their knowledge and skills. This subchapter will provide practical guidance on how to pursue a successful career in cybersecurity, including:

- Identifying relevant certifications and training programs to enhance skills.
- Participating in industry conferences and events to expand professional networks.
- Staying informed about the latest cybersecurity trends, regulations, and best practices.
- Cultivating a proactive and problem-solving mindset to address emerging threats.
- Collaborating with cross-functional teams to build a culture of security awareness within organizations.

By embracing the principles of cybersecurity and information assurance, employees can not only protect their organizations from potential threats but also position themselves as valuable assets within their respective industries. This subchapter aims to equip employees with the necessary knowledge and tools to navigate the path to professional growth in the digital world and contribute to the overall security and success of their organizations.

Remember, in the ever-evolving digital landscape, cybersecurity is not just an option; it is an essential component for every organization's survival.

User Experience and Design

In today's digital world, user experience (UX) and design have become paramount in the success of any product or service. Employers across various industries are recognizing the value of investing in UX and design professionals to enhance their offerings and gain a competitive edge. This subchapter explores the crucial role of UX and design in career growth and provides guidance for professionals seeking to master this field.

User experience refers to how users interact with a product or service, including their emotions, attitudes, and overall satisfaction. Design, on the other hand, focuses on creating visually appealing and functional interfaces that align with user needs and preferences. Together, UX and design have the power to create seamless and enjoyable experiences that keep users coming back for more.

For employers, prioritizing UX and design is essential for staying relevant in a rapidly evolving digital landscape. By investing in these areas, companies can improve customer satisfaction, increase user engagement, and drive business growth. As a result, demand for UX and design professionals is steadily increasing, making it an attractive niche for individuals seeking career growth opportunities.

To excel in this field, professionals must possess a deep understanding of user behavior, design principles, and industry trends. They should be skilled in conducting user research, prototyping, and usability testing to ensure that the end product meets user expectations. Additionally, staying up to date with the latest design software and tools is crucial for delivering high-quality work efficiently.

For those seeking to master UX and design, continuous learning is key. There are numerous online courses, workshops, and certifications available that can help individuals develop and refine their skills. Building a strong portfolio showcasing various projects and demonstrating problem-solving abilities is also essential for standing out in this competitive industry.

Moreover, networking with other professionals in the UX and design community can provide valuable insights and opportunities for collaboration. Attending industry conferences and joining online communities can help expand professional networks and stay abreast of the latest trends and technologies.

In conclusion, UX and design play a pivotal role in career growth in the digital world. Employers are increasingly recognizing the importance of investing in these areas to enhance their offerings and gain a competitive advantage. For professionals seeking to master UX and design, continuous learning, building a strong portfolio, and networking are crucial steps to take. By staying ahead of industry trends and delivering exceptional user experiences, individuals can unlock exciting opportunities and fuel their professional growth.

Project Management in Tech

In today's rapidly evolving digital world, project management plays a pivotal role in the success of tech organizations. As technology continues to advance at an unprecedented pace, businesses need skilled project managers who can navigate complex projects, ensure efficient resource allocation, and deliver high-quality outcomes. This subchapter, titled "Project Management in Tech," aims to provide valuable insights and practical tips for employees in the tech industry to enhance their career growth and excel in project management.

The tech industry is known for its fast-paced nature, where innovation and timelines are critical. Effective project management is the key to ensuring that projects are completed on time, within budget, and meet the desired objectives. This subchapter will delve into the core principles of project management, such as defining project scope, establishing clear goals, and creating a well-defined project plan.

One of the key aspects of project management in the tech industry is the incorporation of agile methodologies. Agile project management enables teams to adapt to changing requirements and deliver incremental value to stakeholders. This subchapter will explore various agile frameworks, including Scrum and Kanban, and provide guidance on implementing them effectively.

Furthermore, this subchapter will delve into the essential skills and qualities required for successful project management in the tech industry. From leadership and communication to problem-solving and risk management, it will provide valuable insights into honing these skills to thrive in a tech project management role.

To support career growth in project management, this subchapter will also discuss the various certifications and professional development opportunities available in the tech industry. From Project Management Professional (PMP) certification to Agile certifications, it will provide guidance on choosing the right certifications and upskilling to enhance career prospects.

Additionally, this subchapter will explore the challenges and best practices in managing remote or distributed teams, a prevalent trend in the tech industry. It will provide strategies for effective collaboration, communication, and project tracking in remote work environments.

By mastering project management in the tech industry, employees can enhance their career growth prospects, become valuable assets to their organizations, and contribute to the successful delivery of innovative digital solutions. This subchapter is a comprehensive guide that equips tech professionals with the knowledge and skills required to excel in project management and navigate the path to professional growth in the digital world.

In conclusion, "Project Management in Tech" is an essential subchapter in the book "Tech Career Mastery: Navigating the Path to Professional Growth in the Digital World." It addresses the niche of career growth and provides employees in the tech industry with valuable insights, practical tips, and strategies to excel in project management, ensuring successful outcomes in an ever-evolving digital landscape.

Tech Sales and Marketing

Tech Sales and Marketing: Accelerating Career Growth in the Digital World

In today's fast-paced and ever-evolving digital landscape, the role of tech sales and marketing professionals has become increasingly crucial. As technology continues to shape businesses across industries, the need for skilled individuals who can effectively promote and sell these technological solutions has skyrocketed. This subchapter aims to provide invaluable insights and strategies to navigate the path to professional growth in the realm of tech sales and marketing.

Tech sales and marketing professionals play a vital role in driving revenue and growth for their organizations. They are responsible for understanding complex technological products and services and translating their benefits into compelling value propositions for potential clients. By leveraging their expertise, these professionals bridge the gap between the technical aspects of the product and the needs and goals of customers.

To excel in this field and achieve career growth, tech sales and marketing professionals must possess a unique blend of skills. They need to have a deep understanding of the products they are selling, as well as the ability to communicate their value effectively. Additionally, they must stay updated with the latest industry trends and continuously adapt their strategies to cater to evolving customer demands.

This subchapter will delve into various aspects of tech sales and marketing, equipping you with the knowledge and tools you need to excel in this competitive field. It will explore effective sales and marketing techniques, how to build and maintain strong customer relationships, and strategies for leveraging technology to enhance your performance.

Furthermore, this subchapter will address the importance of developing a personal brand and establishing thought leadership in the tech sales and marketing space. By positioning yourself as an authority in your niche, you can attract new opportunities and open doors for career advancement.

Whether you are just starting your journey in tech sales and marketing or looking to accelerate your career growth, this subchapter will serve as your roadmap to success. It will provide you with actionable insights, real-life examples, and practical tips to enhance your skills, increase your value proposition, and navigate the ever-changing digital landscape.

Embrace the opportunities that technology offers and embark on a fulfilling and prosperous career in tech sales and marketing. Let this subchapter be your guide as you navigate the path to professional growth in the digital world.

Chapter 3: Building a Strong Foundation for Tech Career Success

Identifying Your Skills and Interests

In today's rapidly evolving digital world, a successful career in the tech industry requires more than just technical proficiency. It demands a deep understanding of your own skills and interests, as well as the ability to align them with the ever-changing demands of the market. This subchapter will guide you through the process of self-discovery, helping you identify your unique set of skills and interests to pave the way for career growth in the tech industry.

To begin this journey, take a moment to reflect on your current role and responsibilities. What aspects of your job do you find most fulfilling? What tasks do you excel at, and which ones do you struggle with? Understanding your strengths and weaknesses is crucial in determining where you can excel and where you may need additional development.

Next, consider the broader landscape of the tech industry. What emerging trends and technologies captivate your interest? Are you passionate about artificial intelligence, cybersecurity, or data analytics? By identifying your areas of interest, you can align your career goals with the niche that resonates with you the most.

Once you have a sense of your skills and interests, it's important to assess their market relevance. Conduct research on job listings and industry reports to buidentify the skills that are in high demand. Look for patterns and identify the gaps between your current skill set and what the market demands. This will help you determine which skills to focus on developing further.

In addition to technical skills, soft skills are equally important for career growth. Assess your communication, leadership, and problem-solving abilities. Identify areas where you can improve and seek out opportunities to strengthen these skills, such as workshops, online courses, or networking events.

Remember that identifying your skills and interests is an ongoing process. As the tech industry continues to evolve, so will your career aspirations. Stay curious, embrace lifelong learning, and be open to exploring new areas of interest. This will not only help you stay relevant in the ever-changing tech landscape but also enable you to adapt and thrive in your professional journey.

In conclusion, identifying your skills and interests is a fundamental step towards achieving career growth in the tech industry. By understanding your strengths, aligning them with your passions, and staying attuned to market demands, you can position yourself for success. So, take the time to reflect, research, and invest in your personal and professional development – the rewards will be well worth it.

Assessing the Educational Requirements

When it comes to career growth in the rapidly evolving digital world, assessing the educational requirements becomes crucial for aspiring professionals. Technology is constantly advancing, and staying ahead of the game requires a deep understanding of the industry and its demands. In this subchapter, we will delve into the importance of assessing educational requirements and how it can pave the way for success in the tech industry.

In today's competitive job market, employers are seeking candidates with the right skill set and educational background. Understanding the educational requirements within your chosen niche is vital for career advancement. Whether you aim to become a data scientist, software engineer, or cybersecurity expert, having the necessary educational foundation will set you apart from the competition.

Assessing educational requirements begins with self-reflection and understanding your career goals. Identify the specific niche or field you wish to pursue and research the educational qualifications typically required for professionals in that area. Look for industry standards, certifications, and advanced degrees that can enhance your expertise and credibility.

While formal education is important, it's equally vital to keep up with the latest industry trends and technologies. The digital world is constantly evolving, and staying knowledgeable about emerging tools and techniques can give you a competitive edge. Seek out continuing education opportunities, online courses, and workshops that can supplement your formal education and keep your skills up to date.

Networking and mentorship can also play a significant role in assessing educational requirements. Connect with professionals in your desired niche and seek guidance on the educational paths that have proven beneficial in their careers. Engage in industry events, conferences, and webinars to expand your network and gain valuable insights into the educational landscape of your chosen field.

In conclusion, assessing educational requirements is a critical step in achieving career growth in the digital world. By understanding the educational qualifications needed within your niche, you can make informed decisions about your educational journey. Whether it's pursuing a formal degree, obtaining industry certifications, or staying updated with the latest trends, investing in your education will position you for success in the tech industry. Remember, continuous learning and adaptability are key to thriving in this ever-changing digital landscape.

Gaining Practical Experience through Internships and Projects

In today's fast-paced and competitive digital world, practical experience is often the key to unlocking career growth opportunities. Employers value candidates who have hands-on experience and can hit the ground running. One of the most effective ways to gain such experience is through internships and projects.

Internships provide an invaluable opportunity to bridge the gap between academic learning and real-world application. They allow you to step into a professional environment and gain exposure to the industry you aspire to work in. During an internship, you will have the chance to work alongside experienced professionals, ask questions, and observe how things are done. This firsthand experience not only enhances your technical skills but also helps you develop essential soft skills, such as communication, teamwork, and problem-solving. Moreover, internships often serve as a stepping stone to full-time employment, as employers often hire interns who have proved their worth during their internship period.

Projects, on the other hand, offer a more flexible and self-directed approach to gaining practical experience. Whether you are a student, a recent graduate, or someone looking to transition into a new career, taking on projects related to your desired field can significantly boost your skillset and marketability. Projects can be personal initiatives, freelance work, or collaborations with other professionals. They allow you to apply your theoretical knowledge to real-world challenges, showcase your abilities, and build a portfolio that demonstrates your expertise.

Both internships and projects offer unique advantages for career growth. They provide opportunities to network with professionals in your chosen field, gain industry insights, and build relationships that can lead to future job opportunities. Additionally, practical experience gained through internships and projects sets you apart from other candidates during the job application process, giving you a competitive edge.

To make the most of internships and projects, it is crucial to approach them with a growth mindset. Be proactive, ask for feedback, and seek opportunities to expand your responsibilities. Treat these experiences as learning opportunities and seize every chance to develop new skills and knowledge. Remember, the skills and connections you gain during internships and projects can be instrumental in shaping your career trajectory.

In conclusion, internships and projects are powerful vehicles for gaining practical experience and accelerating career growth in the digital world. They provide a hands-on learning environment, enhance technical and soft skills, and open doors to new opportunities. Embrace these opportunities, leverage your experiences, and take control of your career journey. With the right mindset and dedication, internships and projects can be transformative stepping stones toward professional success in the ever-evolving tech industry.

Networking and Building Professional Relationships

In today's digital world, where technology is constantly evolving and industries are becoming increasingly interconnected, building a strong professional network and establishing meaningful relationships is more important than ever. In this subchapter, we will delve into the crucial aspects of networking and how it plays a pivotal role in career growth.

Networking is not just about attending events and collecting business cards; it is about cultivating genuine connections with like-minded professionals in your industry. By building a robust network, you can open doors to new opportunities, gain valuable insights, and receive support during challenging times. It is the key to unlocking career growth in the digital landscape.

To effectively build professional relationships, it is essential to approach networking with a strategic mindset. Start by identifying your goals and objectives. Are you looking for mentorship, job opportunities, or simply expanding your knowledge base? Once you have a clear vision of what you want to achieve, you can tailor your networking efforts accordingly.

Take advantage of both offline and online platforms to expand your network. Attend industry conferences, seminars, and workshops where you can meet professionals from various niches and exchange ideas. Utilize social media platforms like LinkedIn to connect with individuals who share similar interests and engage in conversations within professional groups.

When networking, remember that it is a two-way street. Be genuinely interested in others and their work. Actively listen and ask thoughtful questions to establish a meaningful connection. Offer your expertise and support whenever possible, as relationships thrive on reciprocity.

Networking is not limited to professionals in your immediate field. Diversify your connections by reaching out to individuals from different industries and backgrounds. These diverse perspectives can provide fresh insights and open doors to unexpected opportunities.

Additionally, networking should not be limited to moments of need. Cultivate relationships continuously and maintain regular communication with your network. Share valuable content, attend networking events, and offer assistance whenever possible. By nurturing your network consistently, you stay top of mind and strengthen the bond with your contacts.

In conclusion, networking and building professional relationships are essential components of career growth in the digital world. By strategically connecting with professionals, cultivating genuine relationships, and actively participating in industry events, you can unlock new opportunities, gain valuable insights, and accelerate your professional growth. Remember, networking is not just about collecting contacts; it is about fostering meaningful connections that can propel your career to new heights.

Developing a Personal Brand in the Tech Industry

In today's competitive tech industry, having a strong personal brand is crucial for career growth and professional success. Your personal brand is what sets you apart from others in the industry, showcasing your unique skills, expertise, and values. It not only helps you stand out in a crowded field but also opens doors to exciting opportunities and positions you as a thought leader in your niche.

One key aspect of developing a personal brand in the tech industry is to identify your unique strengths and skills. Take some time to reflect on your technical expertise, problem-solving abilities, and any niche areas where you excel. By understanding your strengths, you can focus on developing and showcasing them, positioning yourself as an expert in those areas.

Another important aspect of personal branding in the tech industry is building a strong online presence. In today's digital world, employers and potential clients often turn to the internet to research and evaluate professionals. Create a professional website or portfolio that highlights your achievements, projects, and contributions. Additionally, leverage social media platforms like LinkedIn, Twitter, and GitHub to connect with others in the industry, share valuable insights, and establish your expertise.

Networking is also crucial for personal brand development. Attend tech conferences, industry events, and meetups to connect with like-minded professionals and expand your network. Engage in meaningful conversations, share your knowledge, and demonstrate your passion for the industry. By building relationships with other professionals, you can gain valuable insights, access new opportunities, and establish yourself as a trusted resource in the tech community.

Furthermore, actively contribute to the tech industry by sharing your knowledge and insights. Start a tech blog, write articles for industry publications, or speak at conferences and webinars. By creating valuable content and sharing it with others, you can establish yourself as an authority in your niche and gain recognition as a thought leader.

Lastly, don't underestimate the power of continuous learning and staying updated with the latest trends and advancements in the tech industry. Attend workshops, take online courses, and pursue certifications to enhance your skills and knowledge. By staying ahead of the curve, you position yourself as a valuable asset to employers and clients.

In conclusion, developing a personal brand in the tech industry is essential for career growth and professional success. By identifying your strengths, building a strong online presence, networking, contributing to the industry, and continuously learning, you can establish yourself as a respected professional and open doors to exciting opportunities in the digital world.

Chapter 4: Navigating Job Search and Career Advancement in Tech

Crafting an Effective Tech Resume

In today's fast-paced digital world, the demand for skilled tech professionals is skyrocketing. As technology continues to advance, employers are constantly seeking talented individuals who can contribute to their organizations' success. To stand out in this competitive landscape, it is crucial for tech professionals to create an effective resume that showcases their skills, experience, and potential.

The first step in crafting an effective tech resume is to understand the specific requirements of the job you are applying for. Research the company and the role you are interested in to identify the key skills and qualifications they are seeking. Tailor your resume to highlight your relevant experience and expertise in those areas. This customization will demonstrate your understanding of the company's needs and increase your chances of getting noticed.

When it comes to formatting your resume, simplicity is key. Keep the design clean and professional, using a consistent font and formatting throughout. Use bullet points to make your achievements and skills easily scannable for employers who often have limited time to review resumes. Remember to include your contact information and a professional summary or objective at the beginning of your resume to quickly capture the employer's attention.

In the tech industry, employers value practical skills and experience. Be sure to include a section highlighting your technical expertise, including programming languages, software applications, and any certifications you have obtained. Additionally, showcase your problem-solving abilities and any successful projects you have worked on. Providing concrete examples of your accomplishments will demonstrate your value as a potential employee.

It is also important to highlight your ongoing commitment to professional growth and learning. Include any relevant training courses, workshops, or industry events you have attended. This will show employers that you are dedicated to staying up-to-date with the latest advancements in the field.

Finally, proofread your resume multiple times to ensure it is free from errors and typos. Even the smallest mistake can create a negative impression and potentially cost you an opportunity. Consider seeking feedback from colleagues or friends to ensure your resume is polished and impactful.

Crafting an effective tech resume is essential for career growth in the digital world. By understanding the specific requirements of the job, showcasing your skills and experience, and emphasizing your commitment to professional growth, you can create a compelling resume that will grab the attention of employers and open doors to exciting opportunities in the tech industry.

Mastering the Art of Job Interviews

In today's competitive job market, the ability to ace job interviews is an invaluable skill for professionals seeking career growth in the digital world. Whether you are a seasoned employee looking for a new challenge or a recent graduate embarking on your professional journey, understanding the art of job interviews can significantly increase your chances of securing your dream job. This subchapter of "Tech Career Mastery: Navigating the Path to Professional Growth in the Digital World" aims to equip you with the knowledge and strategies needed to master the interview process.

The first step in mastering job interviews is thorough preparation. Research the company and the role you're applying for, as well as the industry trends and challenges they face. This will allow you to tailor your answers to showcase how your skills and experiences align with their needs. Additionally, practice common interview questions and develop concise and compelling answers that highlight your strengths and achievements.

Furthermore, mastering non-verbal communication is crucial during job interviews. Pay attention to your body language, maintain eye contact, and project confidence through a firm handshake. Dress professionally and appropriately for the industry, as first impressions are often formed within seconds of meeting the interviewer.

During the interview, actively listen to the questions and take your time to answer thoughtfully. Use the STAR method (Situation, Task, Action, Result) to structure your responses to behavioral questions, providing specific examples from your past experiences to demonstrate your skills and problem-solving abilities. Remember to emphasize your ability to adapt to the ever-changing digital world and your willingness to continuously learn and grow.

In addition to your technical skills, employers also value soft skills such as teamwork, communication, and problem-solving. Showcase these skills by providing examples of situations where you successfully collaborated with others or resolved conflicts. Employers are not only looking for technical expertise but also for individuals who can thrive in a dynamic and collaborative work environment.

Lastly, always conclude the interview with thoughtful questions for the interviewer. This demonstrates your genuine interest in the role and the company. Asking about the company culture, growth opportunities, or current projects can leave a lasting impression and show that you are invested in your career growth within the organization.

Mastering the art of job interviews is a continuous journey that requires practice, self-reflection, and refinement. By adequately preparing, leveraging effective communication techniques, and showcasing your skills and experiences, you can confidently navigate the interview process and secure the career growth opportunities you desire in the digital world.

Negotiating Job Offers and Compensation

In today's rapidly evolving digital world, career growth is a top priority for ambitious professionals. As technology continues to revolutionize industries, it is essential for employees to navigate the path to professional growth effectively. One crucial aspect of this journey is negotiating job offers and compensation. This subchapter of "Tech Career Mastery: Navigating the Path to Professional Growth in the Digital World" aims to equip employees with the knowledge and strategies needed to excel in these negotiations.

Negotiating job offers and compensation can be intimidating, but with the right approach, it presents a significant opportunity for career advancement. This subchapter will guide you through the essential steps of negotiation, empowering you to secure the best possible job offer and compensation package.

Firstly, it is essential to understand your worth in the job market. Researching industry standards and salary benchmarks will provide you with valuable insights and enable you to set realistic expectations. Armed with this knowledge, you can confidently articulate your value to potential employers.

Next, we will delve into effective negotiation techniques. From understanding the company's needs and priorities to highlighting your unique skills and experiences, this subchapter will equip you with the tools necessary for successful negotiations. We will explore the art of persuasive communication, emphasizing the importance of active listening and strategic questioning to uncover hidden opportunities.

Moreover, we will discuss the non-monetary aspects of job offers that can significantly impact your career growth. Factors like flexible work arrangements, professional development opportunities, and potential for advancement can be just as valuable as financial compensation. Understanding these variables and negotiating for a comprehensive package will contribute to your overall satisfaction and long-term success.

Additionally, this subchapter will address important considerations such as counteroffers and evaluating multiple job offers. We will provide insights on when and how to negotiate counteroffers and compare offers to make informed decisions aligned with your career goals.

By mastering the art of negotiating job offers and compensation, you will position yourself for accelerated career growth in the digital world. This subchapter of "Tech Career Mastery" will empower you to navigate these crucial negotiations with confidence and maximize your professional potential.

Remember, negotiation is not about winning or losing; it's about reaching a mutually beneficial agreement. With the strategies and insights shared in this subchapter, you will be well-equipped to successfully negotiate job offers and compensation, setting yourself up for a fulfilling and prosperous tech career.

Strategies for Career Advancement in the Tech Industry

In today's rapidly evolving digital world, the tech industry has emerged as a powerhouse of opportunities for professionals seeking career growth. As an employee in this dynamic field, it is crucial to have a clear roadmap and effective strategies in place to ensure continuous advancement. This subchapter will explore key strategies for career advancement in the tech industry, helping you navigate the path to professional growth.

1. Continuous Learning: In the tech industry, staying updated with the latest technologies and trends is paramount. Embrace a growth mindset and invest in continuous learning. Attend workshops, conferences, and webinars, and pursue relevant certifications to enhance your skill set and stay ahead of the curve.

2. Networking: Building a strong professional network can open doors to new opportunities. Attend industry events, join tech communities, and connect with like-minded professionals. Engage in meaningful conversations, share knowledge, and build relationships that can offer invaluable support and guidance throughout your career journey.

3. Mentorship: Seek out experienced professionals who can serve as mentors. A mentor can provide valuable insights, help you navigate challenges, and offer career guidance. Look for opportunities to connect with mentors within your organization or through industry associations.

4. Specialization: Identify a niche within the tech industry that aligns with your interests and strengths. Specializing in a particular area can set you apart from the competition and make you an expert in your field, opening up doors to more challenging and rewarding roles.

5. Proactive Career Planning: Take ownership of your career growth by setting clear goals and creating a roadmap for achieving them. Regularly assess your skills, identify areas for improvement, and seek out opportunities that align with your long-term objectives.

6. Leadership Skills: Developing strong leadership skills is essential for career advancement. Seek opportunities to lead projects or teams, and cultivate skills like effective communication, decision-making, and strategic thinking. These skills will not only enhance your career prospects but also make you a valuable asset to any organization.

7. Embrace Change: The tech industry is known for its rapid pace of change. Embrace new technologies and methodologies, be open to exploring different roles, and be adaptable to change. Embracing change will not only help you stay relevant but also open up new avenues for career growth.

By implementing these strategies, you can position yourself for continuous career advancement in the tech industry. Remember, career growth is a journey, and by investing in your skills, networking, and proactive planning, you can pave the way for a successful and fulfilling career in the digital world.

Continuous Learning and Professional Development

In today's rapidly evolving digital world, staying ahead of the curve is crucial for career growth and success. The ever-changing landscape of technology demands that professionals continuously learn and develop their skills to remain competitive in their respective fields. This subchapter explores the importance of continuous learning and professional development, and provides strategies to navigate the path to professional growth.

Continuous learning is the ongoing process of acquiring new knowledge, skills, and abilities throughout one's career. It is the key to staying relevant and adaptable in the dynamic tech industry. With the rapid pace of technological advancements, professionals must embrace a mindset of lifelong learning to keep up with emerging trends and technologies.

Professional development complements continuous learning by focusing on enhancing specific skills and competencies required for career growth. It involves strategic planning and proactive efforts to acquire new certifications, attend industry conferences, participate in workshops, or pursue advanced degrees. By investing in professional development, employees demonstrate their commitment to personal growth and their willingness to take on new challenges.

To embark on the journey of continuous learning and professional development, employees must first identify their career goals and aspirations. By having a clear vision of where they want to go, they can chart a course that aligns with their objectives. This could involve assessing their current skill set, identifying areas for improvement, and setting achievable milestones to measure progress.

Furthermore, networking plays a crucial role in continuous learning and professional development. Engaging with industry professionals, attending meetups, and joining online communities can provide valuable insights and opportunities for collaboration. These connections can lead to mentorship, partnerships, and exposure to new ideas and perspectives.

In addition, leveraging online learning platforms and resources can enhance the learning experience. Massive Open Online Courses (MOOCs), webinars, and tutorials offer flexible and accessible options to acquire new skills and knowledge. Many of these platforms offer certifications or badges that can be added to a professional's portfolio, further boosting their credibility.

Ultimately, continuous learning and professional development are essential for career growth in the digital world. By embracing a mindset of lifelong learning, setting clear goals, networking, and leveraging online resources, employees can navigate the path to professional growth successfully. Embracing these strategies will not only keep professionals relevant in their current roles but also open doors to new and exciting opportunities in the ever-evolving tech industry.

Chapter 5: Thriving in the Tech Workplace

Understanding Corporate Culture in Tech Companies

In today's fast-paced digital world, tech companies have emerged as some of the most innovative and dynamic organizations. As professionals in the tech industry, it is crucial to have a deep understanding of corporate culture within these companies to navigate the path to career growth successfully. This subchapter will explore the intricacies of corporate culture in tech companies, shedding light on its significance and providing insights into how to thrive in this unique environment.

Corporate culture encompasses the values, beliefs, and behaviors that shape the collective identity of an organization. In tech companies, the culture often revolves around innovation, collaboration, and a passion for cutting-edge technology. Understanding and aligning with these cultural aspects can significantly impact your career trajectory.

One key aspect of corporate culture in tech companies is a relentless pursuit of innovation. These organizations thrive on pushing the boundaries of what is possible, constantly seeking new solutions to complex problems. To succeed in such an environment, it is essential to embrace a growth mindset and actively contribute to the culture of innovation. Taking calculated risks, embracing failure as a learning opportunity, and constantly seeking ways to improve are all valued traits.

Another essential element is collaboration. Tech companies often foster a culture of cross-functional teamwork, recognizing the power of diverse perspectives. Being able to effectively collaborate and communicate with individuals from different backgrounds and areas of expertise is crucial for success. By actively seeking opportunities to collaborate and leveraging the strengths of your team members, you can contribute to a culture of shared success.

Tech companies also place a strong emphasis on professional development and learning. Continuous learning is integral to staying relevant and thriving in the ever-evolving tech industry. By taking advantage of the learning opportunities provided by your organization, such as training programs, conferences, and mentorship, you can demonstrate your commitment to personal growth and development.

Lastly, understanding the importance of work-life balance is crucial in tech companies. While the industry can be demanding, it is essential to prioritize self-care and maintain a healthy work-life balance. Companies that value their employees' well-being often have better retention rates and higher levels of job satisfaction. Setting clear boundaries, practicing self-care, and advocating for work-life balance can contribute to a positive corporate culture.

In conclusion, understanding corporate culture in tech companies is vital for career growth in the digital world. By embracing a culture of innovation, collaboration, continuous learning, and work-life balance, you can position yourself for success within these organizations. Remember, thriving in a tech company goes beyond technical skills – it requires a deep understanding and alignment with the unique cultural aspects that drive these organizations forward.

Effective Communication in a Tech Environment

In today's digital world, effective communication skills have become more important than ever, especially in a tech environment. As technology continues to evolve at a rapid pace, the ability to communicate clearly, efficiently, and professionally is crucial for career growth and success in the tech industry.

One of the key aspects of effective communication in a tech environment is the ability to convey complex technical concepts to both technical and non-technical audiences. As a tech professional, you will often find yourself in situations where you need to explain complex ideas to clients, colleagues, or stakeholders who may not have the same level of technical expertise. Being able to break down complex concepts into simple and understandable terms is a valuable skill that will set you apart from your peers and help you advance in your career.

Additionally, effective communication plays a vital role in collaborating with teams and managing projects in a tech environment. Clear and concise communication ensures that everyone is on the same page, understands their roles and responsibilities, and can work together efficiently and effectively. It helps to eliminate misunderstandings, reduces errors, and fosters a collaborative and productive work environment.

Furthermore, communication skills are essential for building and maintaining professional relationships in the tech industry. Networking is a crucial aspect of career growth, and the ability to communicate effectively allows you to connect with industry professionals, mentors, and potential employers. Strong communication skills help you present yourself confidently, articulate your ideas, and establish rapport with others, opening doors to new opportunities and career advancements.

To improve your communication skills in a tech environment, consider the following strategies:

1. Practice active listening: Actively listen to others, pay attention to their ideas, and ask clarifying questions to ensure understanding.

2. Tailor your communication style: Adapt your communication style to suit different audiences, considering their level of technical knowledge and preferred communication methods.

3. Use visual aids: Utilize visual aids such as diagrams, charts, or presentations to enhance your communication and make complex concepts more accessible.

4. Seek feedback: Regularly seek feedback from colleagues, mentors, or supervisors to identify areas for improvement and continually refine your communication skills.

In conclusion, effective communication is fundamental for career growth in the tech industry. By honing your communication skills, you will be able to convey complex ideas, collaborate effectively, and build strong professional relationships, propelling your career to new heights in the digital world.

Collaboration and Teamwork in Tech Projects

In today's fast-paced digital world, collaboration and teamwork are crucial components for achieving successful outcomes in tech projects. The ability to work effectively with others is not only advantageous for personal career growth but also essential for the overall success of any organization. This subchapter explores the significance of collaboration and teamwork in tech projects and provides valuable insights for employees looking to enhance their professional growth in this ever-evolving industry.

Effective collaboration in tech projects involves leveraging the diverse skills and expertise of team members to achieve common goals. By working together, individuals can harness their collective abilities, creativity, and problem-solving skills to overcome challenges and deliver high-quality results. In the rapidly evolving tech landscape, no single person can possess all the necessary skills and knowledge to tackle complex projects alone. Thus, collaboration becomes the key to success.

One aspect of collaboration in tech projects is effective communication. Clear and concise communication ensures that team members are on the same page and have a shared understanding of project objectives and expectations. A culture of open communication encourages the exchange of ideas, feedback, and constructive criticism, fostering an environment where innovation can thrive.

Another crucial element of collaboration in tech projects is the ability to work harmoniously as a team. Successful teams embrace diversity, appreciate each other's strengths, and understand the importance of individual contributions. By fostering a supportive and inclusive environment, team members feel empowered to share their thoughts and ideas, resulting in enhanced problem-solving capabilities and innovative solutions.

Technology has made remote collaboration a possibility, allowing team members to work together regardless of their physical location. Virtual collaboration tools and platforms have revolutionized the way teams operate, enabling seamless communication, file sharing, and project management. Embracing these technologies can significantly enhance teamwork and productivity in tech projects.

For employees seeking career growth in the tech industry, developing strong collaboration and teamwork skills is essential. Organizations value individuals who can work effectively in teams, demonstrating their ability to contribute to collective success while also fostering a positive work environment. By actively participating in collaborative projects, employees can enhance their skills, expand their professional networks, and gain valuable experience that can propel their careers forward.

In conclusion, collaboration and teamwork are vital components for success in tech projects. By embracing effective communication, fostering a supportive team culture, and leveraging technology to facilitate collaboration, employees can enhance their professional growth and contribute to the overall success of their organizations in the fast-paced digital world.

Managing Time and Priorities in a Fast-Paced Tech Industry

In today's fast-paced tech industry, time management and prioritization skills are crucial for career growth and success. With ever-evolving technologies, tight deadlines, and an ever-increasing workload, it can be challenging to stay on top of everything. However, by adopting effective time management strategies and prioritizing tasks, you can navigate the path to professional growth in the digital world.

One of the first steps in managing time and priorities is understanding your goals and objectives. Take the time to identify what you want to achieve in your career and break it down into smaller, actionable steps. By having a clear vision of your goals, you can prioritize tasks that align with your long-term objectives.

Another key aspect of time management is mastering the art of task prioritization. Not all tasks hold the same level of importance, and it's essential to distinguish between urgent and non-urgent tasks. Urgent tasks require immediate attention and should be tackled first, while non-urgent tasks can be scheduled for later. Additionally, consider the impact and potential consequences of each task to prioritize accordingly.

To effectively manage your time, it's crucial to eliminate distractions. The tech industry is filled with constant notifications, emails, and social media distractions that can hinder productivity. Set specific time blocks for checking emails and notifications, and utilize productivity tools that help you stay focused, such as time-tracking apps or website blockers.

Another helpful time management technique is the Pomodoro Technique. This method involves working in focused bursts of 25 minutes, followed by a short break. By breaking your work into manageable chunks, you can maintain focus and productivity throughout the day.

Additionally, learning to delegate tasks and say no when necessary is essential. You can't do it all, and trying to take on too much will only lead to burnout. Learn to trust your team members and assign tasks that play to their strengths. By delegating responsibilities and saying no to non-essential tasks, you create space for the most critical work.

Finally, don't forget to schedule regular breaks and time for self-care. Taking care of your physical and mental well-being is vital for sustained success in a fast-paced tech industry. Engaging in activities such as exercise, meditation, or hobbies can help recharge your energy and improve focus.

In conclusion, managing time and priorities in a fast-paced tech industry is crucial for career growth. By understanding your goals, prioritizing tasks, eliminating distractions, and practicing self-care, you can navigate the path to professional growth in the digital world. Remember, effective time management is not just about being busy; it's about being productive and achieving meaningful results.

Dealing with Challenges and Conflicts in the Tech Workplace

In today's fast-paced digital world, the tech industry has become a hub of innovation and growth. As technology continues to evolve, so do the challenges and conflicts that arise in the tech workplace. In order to achieve career growth in this competitive industry, it is essential to develop strategies for effectively dealing with these challenges.

One of the major challenges faced by tech professionals is the constant need to stay updated with the latest trends and technologies. With new advancements occurring at a rapid pace, it can be overwhelming to keep up. However, embracing a growth mindset and actively seeking out opportunities for learning and development can help professionals navigate this challenge. Engaging in continuous learning, attending workshops, and participating in industry events can enable individuals to stay ahead of the curve and remain relevant in their field.

Another challenge that often arises in the tech workplace is conflicts among team members. With diverse skill sets and personalities, conflicts are bound to occur. However, it is crucial to address these conflicts in a constructive and respectful manner. Effective communication and active listening skills play a vital role in resolving conflicts. Encouraging open dialogue, fostering a culture of collaboration, and practicing empathy can help create a harmonious work environment.

Furthermore, the tech industry is known for its demanding work culture, which can lead to burnout and stress. In order to maintain career growth, it is essential to prioritize work-life balance and mental well-being. Setting boundaries, taking breaks, and engaging in activities outside of work can help prevent burnout and increase productivity.

Additionally, the tech industry is constantly evolving, which often requires professionals to adapt and learn new skills. Embracing change and being open to acquiring new knowledge is crucial for career growth. Seeking out opportunities to expand one's skill set, such as taking on challenging projects or volunteering for cross-functional teams, can contribute to professional development.

In conclusion, navigating the path to professional growth in the tech industry requires addressing challenges and conflicts head-on. By staying updated with the latest technologies, resolving conflicts in a constructive manner, prioritizing work-life balance, and embracing change, tech professionals can overcome obstacles and achieve success in their careers. Remember, a proactive approach to dealing with challenges will not only contribute to personal growth but also enhance the overall work environment for everyone involved.

Chapter 6: Embracing Leadership and Entrepreneurship in Tech

The Role of Leadership in the Tech Industry

In today's rapidly evolving tech industry, leadership plays a crucial role in driving success and fostering career growth. As technology continues to advance at an unprecedented pace, employers need leaders who can navigate the complexities of the digital world and inspire their teams to achieve their full potential.

One key aspect of leadership in the tech industry is the ability to adapt and embrace change. Technology is constantly evolving, and leaders must be able to stay ahead of the curve and guide their teams through the ever-changing landscape. They must be open to new ideas, willing to take risks, and encourage their employees to embrace innovation. By fostering a culture of continuous learning and improvement, leaders can position their companies and employees for long-term success.

Effective leadership also involves setting clear goals and expectations. In the tech industry, where projects can be complex and require collaboration across multiple teams, leaders must provide clarity and direction. By establishing a shared vision and ensuring everyone understands their role in achieving it, leaders can drive alignment and maximize productivity.

Furthermore, leaders in the tech industry must possess strong communication skills. They must be able to effectively convey their ideas, provide feedback, and build relationships. Communication is crucial for coordinating teams, resolving conflicts, and ensuring that

everyone is on the same page. By fostering a transparent and open communication environment, leaders can build trust and empower their employees to contribute their best work.

Another essential aspect of leadership in the tech industry is the ability to nurture talent and promote career growth. As the industry becomes increasingly competitive, attracting and retaining top talent is critical for success. Leaders must invest in their employees' development, provide opportunities for growth, and empower them to take on new challenges. By doing so, leaders can create a culture of continuous learning, improve employee engagement, and build a strong pipeline of future leaders.

In conclusion, leadership plays a vital role in the tech industry's success and the career growth of its employees. By embracing change, setting clear goals, communicating effectively, and nurturing talent, leaders can navigate the complexities of the digital world and drive their teams towards professional growth. With the right leadership, companies can thrive in the ever-evolving tech industry and create a culture of innovation and success.

Developing Leadership Skills and Mindset

In today's fast-paced and ever-evolving digital world, the importance of developing leadership skills and mindset cannot be overstated. As technology continues to advance and reshape industries, employees are expected to adapt and take charge of their career growth. The ability to lead effectively is no longer limited to those in managerial positions; it has become a necessary skill set for all professionals.

Leadership is not merely a title; it is a mindset that involves taking ownership of one's work, embracing challenges, and inspiring others to achieve their full potential. In the context of career growth, developing leadership skills is essential for individuals who aspire to take on more significant responsibilities, make a lasting impact, and advance in their chosen field.

One of the fundamental aspects of developing leadership skills and mindset is self-awareness. Understanding one's strengths, weaknesses, and areas for improvement is crucial for personal growth and effective leadership. By identifying areas of expertise and areas that require development, employees can proactively seek out opportunities to enhance their skills and knowledge.

Another vital aspect of leadership development is fostering effective communication skills. The ability to communicate clearly and persuasively is essential for conveying ideas, motivating others, and building strong professional relationships. Whether it's presenting ideas to colleagues, negotiating with clients, or leading a team, effective communication plays a pivotal role in career growth.

Additionally, leadership development involves cultivating a growth mindset. Embracing a growth mindset means viewing challenges and

setbacks as opportunities for learning and improvement. By adopting this mindset, employees can develop resilience, embrace change, and continuously adapt to the dynamic digital landscape.

To further develop leadership skills and mindset, employees can seek out opportunities for professional development. This may include attending workshops, conferences, or enrolling in leadership training programs. Additionally, seeking mentorship from experienced professionals can provide valuable guidance and insights.

Furthermore, employees can engage in activities that promote their leadership skills, such as taking on leadership roles within their organizations or participating in cross-functional projects. These experiences provide opportunities to practice leadership skills, gain valuable experience, and demonstrate capabilities to potential employers.

In conclusion, developing leadership skills and mindset is critical for career growth in the digital world. It requires self-awareness, effective communication, a growth mindset, and a commitment to continuous learning. By actively seeking opportunities for development, employees can position themselves for success, forge their path to professional growth, and make a lasting impact in their chosen field.

Transitioning from Employee to Tech Entrepreneur

In today's rapidly evolving digital world, the traditional career path of being an employee is no longer the only option available. The rise of technology has opened up a whole new world of opportunities for those who are willing to take the leap and become tech entrepreneurs. This subchapter will guide you through the process of transitioning from being an employee to becoming a tech entrepreneur, helping you navigate the path to professional growth in the digital world.

The decision to transition from being an employee to a tech entrepreneur is not one to be taken lightly. It requires careful consideration, planning, and a willingness to embrace uncertainty. However, the rewards can be immense – from the freedom to be your own boss to the potential for unlimited financial growth.

The first step in transitioning to tech entrepreneurship is to assess your skills and passions. What are you good at? What are you passionate about? Identifying your strengths and interests will help you determine the type of tech startup you want to build. It could be an app development company, a software consultancy, or even an e-commerce business. Choose something that aligns with your skills and interests to increase your chances of success.

Next, you need to develop a solid business plan. This includes defining your target market, understanding your competition, and outlining your unique selling proposition. A comprehensive business plan will serve as your roadmap and will help you secure funding if needed.

Networking is crucial when transitioning to tech entrepreneurship. Connect with like-minded individuals, attend industry events, and join online communities to expand your network. Surrounding

Chapter Four

Delegate to Achieve

Amid the variety of daily demands and pressures, leaders often encounter stumbling blocks that hinder the delegation process. Waiting too late to delegate, over-delegating tasks, and succumbing to the allure of micromanagement stand as common pitfalls that can undermine the fabric of teamwork and productivity. When leaders delay delegation, they risk shouldering an unsustainable burden, hampering their ability to innovate and lead effectively. Over-delegating, on the other hand, can overwhelm team members, leading to diminished morale and a loss of cohesion within the organization. Meanwhile, the pervasive shadow of micromanagement stifles creativity, erodes trust, and impedes the autonomy necessary for individual and team growth.

In this chapter, we explore these challenges, unraveling the complexities of effective delegation and offering insights into overcoming its pitfalls. By delving into the detrimental effects of delayed delegation, over-delegation, and micromanagement, we equip leaders with the knowledge and strategies needed to cultivate a more balanced and impactful approach to delegation. Through this exploration, we aim to illuminate the path toward fostering a culture of trust, empowerment, and collaboration within organizations, ultimately driving greater efficiency and success.

The Dangers of Not Delegating

Sometimes, the reluctance or failure to delegate can pose significant risks to both individuals and organizations. Let's explore the multifaceted dangers associated with not delegating effectively:

- **Burnout and Feeling Overwhelmed**: Leaders who shoulder all responsibilities without delegation often find themselves overwhelmed by the sheer volume of tasks. This relentless burden can lead to burnout, a state of emotional, mental, and physical exhaustion, jeopardizing not only the leader's well-being but also their ability to lead effectively. Moreover, burnout can have ripple effects throughout the organization, impacting team morale and productivity.

- **Missed Opportunities for Growth**: Delegation isn't just about offloading tasks; it's also about nurturing talent and fostering growth within the team. When leaders fail to delegate, they deprive their team members of opportunities to develop new skills, broaden their experience, and take on greater responsibilities. This stagnant environment not only stifles individual growth but also hampers organizational agility and innovation.

- **Micromanagement and Lack of Trust**: In the absence of effective delegation, leaders may resort to micromanagement, closely scrutinizing every aspect of tasks and stifling their team members' autonomy. This micromanagement erodes trust within the team, creating a culture of fear and resentment. Team members feel disempowered and demotivated, leading to disengagement and decreased performance.

- **Limited Capacity for Strategic Thinking**: Leaders who are bogged down by day-to-day tasks have little time or mental bandwidth for strategic thinking and long-term planning. Without effective delegation, leaders become mired in the minutiae of daily operations, unable to lift their gaze to the horizon and anticipate future challenges and opportunities. This shortsightedness can impede organizational growth and hinder competitive advantage.

- **Slow Decision-Making and Inefficiency**: When leaders fail to delegate, decision-making processes become centralized and bottlenecked. Every decision must pass through the leader, leading to delays and inefficiencies. Moreover, without delegation, leaders may find themselves mired in low-level tasks that could be better handled by others, wasting precious time and resources that could be allocated to higher-value activities.

Avoid Mistakes When Delegating

Waiting Too Late

Waiting too late to delegate is a common trap that leaders and entrepreneurs often find themselves ensnared in. It's a subtle pitfall, one that can creep up gradually as responsibilities pile up and the pressure to perform mounts. At first glance, delaying delegation may seem like a prudent decision, driven by a desire to maintain control and ensure that tasks are handled with precision. However, beneath the surface lies a host of detrimental consequences that can impede growth, hinder productivity, and, ultimately, undermine success.

One of the primary signs of waiting too late to delegate is a mounting workload that surpasses manageable levels. As tasks accumulate and deadlines loom, it's tempting to adopt a "just one more thing" mentality, believing that taking on additional responsibilities is a temporary sacrifice in pursuit of long-term gains. However, this mindset often leads to a vicious cycle of overextension, where the leader becomes increasingly overwhelmed and their performance begins to suffer.

Delaying delegation can result in missed opportunities for growth and development, both for the leader and their team. By shouldering the burden of every task themselves, leaders inadvertently deprive their team members of the chance to expand their skills, take on new challenges, and contribute meaningfully to the organization's success. This not only stifles individual growth but also hampers the collective progress of the team and the company as a whole.

Furthermore, waiting too late to delegate can have a detrimental impact on the leader's well-being and mental health. The relentless pressure of trying to juggle an ever-expanding workload can lead to burnout, stress, and a diminished capacity to perform effectively. This not only affects the leader's personal life but also reverberates throughout the organization, creating a culture of exhaustion and dissatisfaction.

In addition to the immediate consequences, delaying delegation can also have long-term implications for the organization's growth and sustainability. By hoarding tasks and responsibilities, leaders inhibit the scalability of their business and limit its potential for expansion. Without the ability to delegate effectively, the organization becomes bottlenecked, unable to adapt to changing market conditions or capitalize on emerging opportunities.

Leaders can avoid the trap of waiting too late to delegate. It all starts with a shift in mindset—one that recognizes delegation not as a sign of weakness or relinquishment of control but as a strategic decision that enables growth and fosters success. Leaders must learn to trust their team members, empower them to take on responsibilities, and provide the necessary support and guidance to ensure their success.

Leaders must cultivate a proactive approach to delegation, anticipating future needs and identifying opportunities to offload tasks before they become overwhelming. This requires a keen awareness of one's own limitations and a willingness to relinquish control in favor of long-term sustainability and growth.

Over-Delegating

Over-delegating is a phenomenon often seen in leaders and managers who are eager to lighten their workload or simply unsure of how to effectively manage their responsibilities. It occurs when individuals assign too many tasks to their team members without considering their capacity or workload. While delegation is essential for effective leadership and team productivity, over-delegating can have detrimental effects on both individuals and the organization as a whole.

One of the consequences of over-delegating is the risk of overwhelming team members and causing burnout. When individuals are inundated with tasks beyond their capacity, they may struggle to prioritize effectively, meet deadlines, or maintain the quality of their work. This not only diminishes their performance but also erodes morale and motivation, leading to increased absenteeism, turnover, and disengagement within the team.

Moreover, over-delegating can result in a lack of cohesion and accountability within the team. When responsibilities are distributed haphazardly or without clear guidance, team members may feel disconnected from the overall vision and objectives of the organization. This can lead to confusion, resentment, and a sense of disempowerment, ultimately hindering collaboration and productivity.

Over-delegating can also have negative implications for the leader's reputation and credibility. When leaders indiscriminately offload tasks onto their team members without providing adequate support or supervision, they risk being perceived as indifferent or incompetent. This can damage trust and undermine the leader's authority, making it difficult to effectively lead and inspire their team.

In addition to the immediate impacts on team dynamics and individual well-being, over-delegating can also have long-term consequences for the organization's success and sustainability. When tasks are assigned without regard for team members' skills, expertise, or workload, there is a heightened risk of errors, delays, and missed opportunities. This can result in decreased efficiency, customer dissatisfaction, and ultimately, damage to the organization's reputation and bottom line.

So how can leaders avoid the pitfalls of over-delegating? It starts with a thoughtful and strategic approach to delegation that takes into account the needs, strengths, and limitations of both the leader and their team members. Before assigning tasks, leaders should carefully assess each team member's workload, skills, and capacity to ensure that they are not overwhelmed. Moreover, leaders should provide clear expectations, guidance, and support to empower their team members to succeed.

Leaders should foster open communication and collaboration within the team, encouraging feedback and dialogue to ensure that tasks are delegated effectively and equitably. By creating a culture of trust, accountability, and empowerment, leaders can mitigate the risks of over-delegating and maximize the potential of their team to achieve success.

Micromanaging

Micromanaging is a leadership style characterized by excessive control, close supervision, and a tendency to oversee every detail of tasks and projects. While the intention behind micromanagement may stem from a desire to ensure quality and efficiency, its effects can be profoundly negative, leading to decreased morale, stifled creativity, and diminished productivity within the team.

One of the most detrimental consequences of micromanaging is the erosion of trust between the leader and their team members. When individuals feel constantly scrutinized and second-guessed, they may begin to question their own abilities and judgment, leading to a loss of confidence and autonomy. This can create a toxic atmosphere of resentment and defensiveness, hindering open communication and collaboration within the team.

Moreover, micromanaging can have a demoralizing effect on team members, sapping their motivation and enthusiasm for their work. When individuals feel that their contributions are not valued or respected, they may become disengaged and apathetic, leading to decreased productivity and job satisfaction. This not only affects individual performance but also undermines the overall effectiveness and success of the team.

Furthermore, micromanagement stifles creativity and innovation within the team, as individuals may be hesitant to take risks or think outside the box for fear of being criticized or reprimanded. When leaders dictate every aspect of a project or task, they limit the potential for new ideas and fresh perspectives, ultimately hindering the organization's ability to adapt and thrive in a rapidly changing environment.

In addition to its negative effects on team dynamics and individual well-being, micromanaging can also have practical implications for the organization's success and sustainability. When leaders devote excessive time and energy to overseeing minute details, they may neglect larger strategic priorities or fail to allocate resources effectively. This can result in missed opportunities, delayed projects, and ultimately, damage to the organization's reputation and bottom line.

So, one of the solutions to avoid the pitfalls of micromanaging and fostering a more positive and productive work environment? It begins with recognizing the signs of micromanagement and understanding its underlying causes. Leaders should reflect on their own motivations and tendencies, acknowledging any fears or insecurities that may be driving their need for control.

Basically, leaders should strive to develop a more hands-off approach, empowering their team members to take ownership of their work and make decisions independently. This involves setting clear expectations and goals, providing support and guidance when needed, and trusting individuals to execute tasks effectively.

Delegate with Purpose

Effective delegation is not merely about offloading tasks; it's about harnessing the collective strength of a team to achieve shared objectives. To delegate with purpose is to approach the process strategically, align tasks with organizational goals, maximize individual strengths, and foster a culture of collaboration and innovation. In this exploration, we delve into the principles of purposeful delegation, uncovering its transformative potential for leaders and organizations alike.

At the heart of purposeful delegation lies a clear understanding of organizational goals and priorities. Leaders must first identify the strategic objectives that drive their organization forward, whether it's increasing revenue, enhancing customer satisfaction, or fostering innovation. With these goals in mind, leaders can then assess the tasks and responsibilities that contribute directly to their achievement. By delegating tasks that align with strategic objectives, leaders ensure that every action taken by their team moves the organization closer to its desired outcomes.

yourself with supportive and knowledgeable individuals can provide invaluable insights and help you open doors to new opportunities.

As an employee, you were used to receiving a steady paycheck. However, in the world of tech entrepreneurship, financial stability may not come as easily in the beginning. It is important to have a financial plan in place to cover your living expenses during the initial stages of your startup. This could include saving money, seeking a loan, or even finding a part-time job to supplement your income.

Lastly, embrace a growth mindset. Being a tech entrepreneur requires continuous learning, adaptability, and resilience. Embrace failure as a learning opportunity and constantly seek ways to improve your skills and business strategies.

Transitioning from being an employee to a tech entrepreneur is an exciting and challenging journey. By assessing your skills and passions, developing a solid business plan, building a strong network, and adopting a growth mindset, you can navigate the path to professional growth in the digital world and unlock the limitless potential of tech entrepreneurship.

Building and Scaling Tech Startups

In today's fast-paced and ever-evolving digital world, tech startups have become a driving force for innovation and growth. These startups have the potential to disrupt industries, create new markets, and transform our lives. However, building and scaling a tech startup is no easy task. It requires a combination of technical expertise, business acumen, and relentless dedication.

This subchapter aims to provide valuable insights and guidance to employees looking to embark on the exciting journey of building and scaling a tech startup. Whether you are a budding entrepreneur or an aspiring intrapreneur within an existing organization, this content will help you navigate the path to professional growth in the digital world.

Building a tech startup involves several key steps. We will discuss the importance of identifying a problem worth solving and how to develop a unique value proposition. We will delve into the process of validating your idea, conducting market research, and building a minimum viable product (MVP). We will also explore the significance of assembling a talented team, securing funding, and creating a strong company culture.

Scaling a tech startup is an entirely different challenge. We will explore various strategies for growth, such as expanding into new markets, scaling your technology infrastructure, and hiring the right people to support your vision. We will discuss the importance of customer acquisition and retention, as well as effective marketing and sales strategies. Additionally, we will address the need for continuous innovation and adapting to evolving market trends.

Throughout this subchapter, we will draw on real-life case studies and insights from successful tech entrepreneurs who have navigated the challenging path of building and scaling startups. Their experiences and lessons learned will provide you with practical guidance and inspiration.

By the end of this subchapter, you will have a comprehensive understanding of the key principles, strategies, and best practices for building and scaling tech startups. You will be equipped with the knowledge and tools necessary to drive career growth within the startup ecosystem, whether as a founder, team member, or intrapreneur. Embrace the opportunities and challenges that lie ahead, and embark on a fulfilling journey of building and scaling tech startups in the digital world.

Navigating the Challenges and Rewards of Entrepreneurship in Tech

Entrepreneurship in the tech industry offers a unique set of challenges and rewards. As technology continues to evolve at an unprecedented pace, the opportunities for career growth in this field are abundant. However, it is important to be aware of the hurdles that come with starting your own tech venture. In this subchapter, we will explore the various challenges and rewards that entrepreneurs face in the tech industry, providing valuable insights for employees looking to embark on their own entrepreneurial journey.

One of the key challenges faced by tech entrepreneurs is the constantly changing landscape of technology. Staying updated with the latest trends and innovations is crucial for success. The rapid pace of change can be overwhelming, making it essential to develop a mindset that embraces continuous learning and adaptation. Entrepreneurs need to be prepared to invest time and effort into research, networking, and attending industry events to stay ahead of the curve.

Another challenge is the fierce competition in the tech market. With numerous startups emerging every day, it's crucial to have a unique value proposition and a solid business plan. Differentiating oneself from competitors is essential to attract customers and secure funding. Building a strong network and establishing strategic partnerships can help overcome this challenge by creating opportunities for collaboration and growth.

Despite the challenges, the rewards of entrepreneurship in tech are numerous. One of the most significant rewards is the opportunity to make an impact. As an entrepreneur, you have the power to solve real-life problems, disrupt industries, and shape the future. The satisfaction of seeing your ideas come to life and positively impacting people's lives is unparalleled.

Financial rewards are also a major motivator for many tech entrepreneurs. The tech industry has immense potential for profitability, with successful startups often experiencing exponential growth. However, it's important to maintain a balance between financial goals and the passion for innovation. Prioritizing long-term sustainability and ethical business practices will contribute to a more fulfilling entrepreneurial journey.

Lastly, entrepreneurship in tech offers unmatched career growth potential. Starting your own venture allows you to take charge of your professional development, explore new areas of interest, and gain diverse skill sets. It also provides the flexibility to work on projects that align with your personal values and beliefs.

In conclusion, entrepreneurship in the tech industry presents both challenges and rewards. Navigating this path requires a growth mindset, adaptability, and a willingness to embrace change. However, the potential for making a significant impact, financial success, and personal growth makes it an enticing option for those seeking career growth in the digital world. By understanding and preparing for the challenges ahead, employees can embark on a rewarding entrepreneurial journey in the tech industry.

Chapter 7: Thriving in the Digital Age: Future Trends and Opportunities

Emerging Technologies and Their Impact on Tech Careers

In today's rapidly evolving digital landscape, emerging technologies have a profound impact on the world of tech careers. As technology continues to advance at an unprecedented rate, it is crucial for professionals in the tech industry to stay ahead of the curve and adapt to the changing demands of their niches. This subchapter explores the various emerging technologies that are reshaping the tech industry and their implications for career growth.

Artificial Intelligence (AI) is one of the most transformative technologies of our time. From machine learning algorithms to natural language processing, AI is revolutionizing how businesses operate. Tech professionals who possess skills in AI development, data analysis, and algorithm design are highly sought after in today's job market. As AI becomes more prevalent, it is essential for employees to upskill and embrace this technology to remain relevant in their careers.

The Internet of Things (IoT) is another game-changing technology that is shaping the tech industry. IoT refers to the network of physical devices, vehicles, appliances, and other objects embedded with sensors and software, enabling them to connect and exchange data. As the IoT ecosystem expands, tech professionals who can design, develop, and manage IoT systems will be in high demand. In addition, the rise of IoT brings new career opportunities in cybersecurity, data analytics, and cloud computing.

Blockchain technology, initially popularized by cryptocurrencies like Bitcoin, has far-reaching implications beyond digital currencies. Blockchain offers secure and transparent data storage and transmission, making it an ideal solution for industries like finance, supply chain management, and healthcare. Professionals with blockchain expertise are poised to have a significant impact on these industries, and the demand for their skills is expected to grow exponentially in the coming years.

Virtual Reality (VR) and Augmented Reality (AR) are rapidly gaining traction in various sectors, including gaming, entertainment, healthcare, and education. Tech professionals who can develop immersive experiences, design user interfaces, and create interactive content for VR and AR platforms are in high demand. The potential for career growth in these fields is vast, as the applications of VR and AR technology continue to expand.

In conclusion, emerging technologies have a transformative impact on tech careers. To thrive in the fast-paced digital world, employees must embrace these technologies, acquire new skills, and adapt to the changing demands of their niches. The ability to stay ahead of the curve and leverage emerging technologies will not only lead to career growth but also open up exciting opportunities in the ever-evolving tech industry.

The Rise of Artificial Intelligence and Machine Learning

In today's digital world, the rapid advancement of technology is reshaping industries and transforming the way we work. One of the most significant developments is the rise of artificial intelligence (AI) and machine learning (ML). These emerging technologies have the potential to revolutionize various sectors and pave the way for new career opportunities.

AI refers to the simulation of human intelligence in machines, enabling them to perform tasks that typically require human intelligence, such as speech recognition, problem-solving, and decision-making. ML, on the other hand, is a subset of AI that focuses on algorithms and statistical models that allow machines to learn and improve from experience without being explicitly programmed.

The integration of AI and ML into various aspects of our lives is already evident. From virtual assistants like Siri and Alexa to autonomous vehicles, these technologies are becoming increasingly commonplace. In addition to consumer applications, AI and ML are also driving innovation across industries such as healthcare, finance, manufacturing, and transportation.

For employees seeking career growth, understanding and embracing AI and ML is crucial. These technologies are creating new job roles and demanding a new set of skills. Professionals who can adapt and leverage AI and ML will likely have a competitive edge in the job market.

While some may have concerns about AI and ML replacing human jobs, the reality is that these technologies are more likely to augment human capabilities rather than replace them entirely. In fact, AI and

ML are expected to create more jobs than they eliminate. However, the nature of these jobs will change, requiring individuals to upskill and reskill to remain relevant.

To thrive in the age of AI and ML, employees should consider acquiring skills such as data analysis, programming, and critical thinking. These skills will enable them to work alongside AI and ML systems, leveraging their power to enhance decision-making and problem-solving.

Moreover, employees should embrace a growth mindset and be open to continuous learning. The field of AI and ML is evolving rapidly, and staying up to date with the latest advancements is essential for career growth. Engaging in online courses, attending workshops, and participating in industry conferences can provide valuable insights and networking opportunities.

In conclusion, the rise of AI and ML presents exciting possibilities for career growth. By understanding and embracing these technologies, employees can position themselves for success in the digital world. With the right skills and a growth mindset, individuals can navigate the ever-changing landscape of technology and embark on a path of professional growth.

The Importance of Ethical Tech Practices

In today's rapidly evolving digital landscape, ethical tech practices have become more important than ever before. As technology continues to shape our world, it is crucial for employees to understand the significance of ethical behavior in their tech careers. This subchapter will explore why ethical tech practices are essential for career growth and success in the digital world.

First and foremost, ethical tech practices build trust. In an era where data breaches and privacy concerns dominate headlines, consumers and businesses alike are becoming increasingly cautious about trusting technology. By adhering to ethical principles, employees can demonstrate their commitment to responsible and secure tech practices, earning the trust of their clients, employers, and colleagues. Trust is the foundation of any successful career, and ethical tech practices play a significant role in establishing and maintaining it.

Moreover, ethical tech practices ensure compliance with legal and regulatory frameworks. With the introduction of stringent data protection laws, such as the General Data Protection Regulation (GDPR), organizations are legally obligated to handle sensitive information responsibly. By following ethical guidelines, employees can help their companies avoid legal complications, penalties, and reputational damage. Compliance with these regulations is not only a legal requirement but also a competitive advantage for organizations in the digital world.

Additionally, ethical tech practices foster innovation. By considering the ethical implications of their work, employees can identify potential risks and find creative solutions to address them. Ethical thinking encourages employees to prioritize the well-being of individuals and society as a whole, leading to the development of more thoughtful and impactful tech solutions. Ethical tech practices can fuel innovation by inspiring employees to leverage their skills and expertise to make a positive difference in the world.

Lastly, ethical tech practices contribute to personal and professional growth. In an industry driven by rapid technological advancements, staying up-to-date with ethical standards is essential for career longevity. By continuously expanding their knowledge and skills in ethical tech practices, employees can position themselves as valuable assets within the organization. Employers value employees who prioritize ethical considerations, as they bring a sense of responsibility and integrity to their work. Moreover, individuals who prioritize ethical practices tend to have a stronger sense of purpose and fulfillment in their careers, leading to higher job satisfaction and long-term success.

In conclusion, ethical tech practices are indispensable for career growth and success in the digital world. By embracing ethical principles, employees can build trust, ensure compliance, foster innovation, and enhance their personal and professional growth. In an industry that is constantly evolving, ethical tech practices are not only the right thing to do but also a strategic advantage for employees seeking to thrive in their tech careers.

Remote Work and Digital Nomadism in the Tech Industry

In today's digital landscape, the tech industry has experienced a significant shift in the way professionals work. The rise of remote work and digital nomadism has opened up a whole new realm of possibilities for career growth and development. This subchapter delves into the world of remote work and digital nomadism in the tech industry, providing valuable insights and guidance for employers who are looking to navigate this evolving landscape.

Remote work has become increasingly popular in recent years, and for good reason. It offers numerous benefits for both employers and employees. For employers, remote work allows for access to a global talent pool, enabling them to hire the best individuals regardless of their geographical location. This not only increases the diversity of the workforce but also fosters innovation and creativity. Additionally, remote work often leads to increased productivity and employee satisfaction, as it provides individuals with the flexibility to work in an environment where they can thrive.

Digital nomadism takes remote work to another level. It refers to individuals who choose to work remotely while constantly traveling, often exploring different cities or even countries. This lifestyle appeals to many tech professionals who crave adventure, flexibility, and the opportunity to immerse themselves in new cultures. Digital nomads are known for their ability to balance work and leisure, leveraging technology to stay connected and productive while on the move.

For employers, embracing remote work and digital nomadism within the tech industry can be a game-changer. By offering flexible work arrangements, companies can attract top talent and retain their best employees. Embracing this shift also allows for cost savings in terms of office space and overhead expenses. However, it is crucial for employers to establish clear guidelines and communication strategies to ensure smooth collaboration and project management among remote teams.

In this subchapter, we will explore the benefits and challenges of remote work and digital nomadism in the tech industry. We will discuss strategies for effective remote team management, tools and technologies that facilitate remote collaboration, and how to foster a strong company culture in a remote work environment. Additionally, we will provide tips for tech professionals who aspire to become digital nomads, including advice on finding remote job opportunities and maintaining work-life balance while on the road.

By embracing remote work and digital nomadism, employers can tap into the immense talent pool available worldwide and create a thriving work environment that promotes career growth and professional development in the tech industry.

Adapting to Constant Change in the Digital World

In today's fast-paced and ever-evolving digital world, the ability to adapt to constant change is crucial for career growth and professional success. As technology continues to advance at an unprecedented rate, those who can embrace and navigate these changes will be best positioned to thrive in their careers.

Adapting to constant change requires a mindset shift. Instead of fearing or resisting change, employees must learn to embrace it as an opportunity for growth and development. The digital landscape is continuously evolving, and those who can adapt their skills and knowledge accordingly will have a competitive edge in the job market.

One of the key ways to adapt to constant change is through continuous learning. In the digital world, new technologies and tools emerge regularly, and staying up-to-date with these advancements is essential. Employers should encourage and support their employees in pursuing ongoing education and training opportunities. This could include attending conferences, enrolling in online courses, or participating in workshops and webinars.

Another vital aspect of adapting to constant change is being open to new ideas and perspectives. In the digital world, innovation is often driven by collaboration and diverse viewpoints. By fostering a culture of openness and inclusivity, employers can create an environment where employees feel comfortable sharing their ideas and experimenting with new approaches. This not only promotes adaptability but also fosters creativity and innovation.

Flexibility is also key to adapting to constant change. As the digital world evolves, so do job roles and responsibilities. Employees must be willing to take on new tasks and adapt their skill sets to meet the changing demands of their roles. This may involve acquiring new skills or even transitioning into entirely new career paths. Being open to change and willing to step outside of one's comfort zone can lead to exciting opportunities for career growth.

Finally, networking and building connections within the digital industry can also aid in adapting to constant change. By engaging with professionals in similar niches, employees can stay informed about industry trends and gain valuable insights into emerging technologies. Networking can also provide access to mentorship opportunities, allowing employees to learn from experienced individuals who have successfully navigated the digital world.

In conclusion, adapting to constant change in the digital world is vital for career growth. By embracing change, continuously learning, being open to new ideas, remaining flexible, and building a strong network, employees can position themselves for success in this ever-evolving landscape. With the right mindset and a commitment to growth, individuals can thrive and excel in their digital careers.

Chapter 8: Balancing Personal and Professional Growth in the Tech Industry

Maintaining Work-Life Balance in a Demanding Tech Career

In today's fast-paced digital world, a demanding tech career can easily consume your life if you let it. Long hours, constant connectivity, and the pressure to stay ahead of the game can leave you feeling overwhelmed and burnt out. However, it is crucial to prioritize work-life balance to ensure sustainable career growth and personal well-being.

First and foremost, it is essential to set clear boundaries between work and personal life. Establishing a designated work area and adhering to a set schedule can help create a separation between your professional and personal life. By defining specific work hours and sticking to them, you can avoid overworking and create time for activities outside of work.

Additionally, learning to manage your time effectively is key to maintaining work-life balance. Prioritize your tasks and allocate specific time slots for each activity. Utilize time management techniques such as time blocking or the Pomodoro Technique to enhance productivity and avoid burnout. By efficiently managing your time, you can accomplish your professional goals while still having time for your personal life.

It's important to remember that downtime is vital for your mental and physical well-being. Make sure to schedule regular breaks throughout the day to recharge and relax. Engage in activities that help you unwind, such as going for a walk, practicing mindfulness or meditation, or pursuing hobbies outside of work. Taking time for yourself will help you stay focused and energized in your career.

Furthermore, communication is crucial in maintaining work-life balance. Clearly communicate your boundaries and expectations to your colleagues and superiors. Let them know when you are not available outside of work hours unless it is an emergency. By setting these expectations early on, you can prevent work from encroaching on your personal time.

Lastly, remember to prioritize self-care. Engage in activities that promote physical and mental well-being, such as regular exercise, eating a balanced diet, getting enough sleep, and spending time with loved ones. Taking care of yourself outside of work will allow you to show up as your best self in your career.

In conclusion, maintaining work-life balance in a demanding tech career is essential for both personal well-being and career growth. By setting boundaries, managing time effectively, taking breaks, communicating expectations, and prioritizing self-care, you can navigate the challenges of the digital world while still enjoying a fulfilling personal life. Remember, work should enhance your life, not consume it.

Managing Stress and Avoiding Burnout

In today's fast-paced digital world, stress and burnout have become common challenges that professionals face in their tech careers. As employees, it is crucial to prioritize managing stress and avoiding burnout to ensure long-term career growth and overall well-being.

Stress can arise from various sources, such as heavy workloads, tight deadlines, constant technological advancements, and the pressure to constantly perform at a high level. However, by implementing effective strategies, individuals can effectively manage stress and prevent it from escalating into burnout.

One of the key ways to manage stress is by practicing self-care. This involves prioritizing physical, mental, and emotional well-being. Engaging in regular exercise, getting enough sleep, and maintaining a healthy diet can significantly reduce stress levels. Additionally, taking breaks and engaging in activities that bring joy and relaxation, such as hobbies or spending time with loved ones, can help alleviate stress.

Another important aspect of managing stress is setting boundaries. In the digital world, it is easy to feel constantly connected and always available. However, it is essential to establish boundaries between work and personal life to prevent burnout. Setting specific working hours, avoiding checking work-related emails or messages outside of those hours, and learning to say no when necessary can help maintain a healthy work-life balance.

Furthermore, effective time management is crucial for stress management. Prioritizing tasks, breaking them down into manageable parts, and creating a realistic schedule can help alleviate the feeling of being overwhelmed. Learning to delegate and ask for help when needed is also important for preventing burnout and maintaining productivity.

Additionally, building a support system is vital for managing stress and avoiding burnout. Connecting with colleagues, mentors, or joining professional communities can provide valuable support, guidance, and a sense of belonging. Sharing experiences, seeking advice, and learning from others' perspectives can help individuals navigate the challenges of their tech careers while reducing stress levels.

Lastly, practicing mindfulness and stress reduction techniques, such as meditation or deep breathing exercises, can help individuals stay present, focused, and grounded amidst the fast-paced nature of the digital world. These techniques can also enhance resilience and enable professionals to bounce back from stressful situations more effectively.

By implementing these strategies and prioritizing self-care, setting boundaries, managing time effectively, building a support system, and practicing mindfulness, employees can successfully manage stress and avoid burnout. This holistic approach to stress management will not only contribute to career growth but also lead to a healthier and more fulfilling professional journey in the digital world.

Building Healthy Habits for Long-Term Success in Tech

In the rapidly evolving world of technology, building healthy habits is crucial for long-term success and career growth. With constant advancements and changing trends, it is essential for tech professionals to stay ahead of the curve and adapt to new challenges. This subchapter will explore some key habits that can help employees in the tech industry navigate their career paths and achieve professional growth in the digital world.

First and foremost, maintaining a continuous learning mindset is paramount. Technology is constantly evolving, and professionals need to stay updated with the latest trends, tools, and techniques. Embracing a growth mindset allows individuals to seek out new opportunities for learning and development, whether it's attending conferences, taking online courses, or joining professional communities. By actively seeking knowledge, tech employees can enhance their skill set and remain relevant in a rapidly changing industry.

Another crucial habit for long-term success in tech is effective time management. With the fast-paced nature of the industry, managing time efficiently is essential to meet deadlines and deliver quality work. Prioritizing tasks, setting realistic goals, and minimizing distractions are all important aspects of effective time management. By adopting strategies such as the Pomodoro Technique or using productivity tools, tech professionals can optimize their work hours and achieve more in less time.

Additionally, building strong professional relationships is vital for career growth in the tech industry. Networking with colleagues, industry experts, and mentors can provide valuable insights, opportunities, and support. Actively participating in professional events, joining relevant industry groups, and leveraging social media platforms can help employees expand their network and tap into new avenues for career advancement.

Lastly, maintaining a healthy work-life balance is crucial for long-term success and overall well-being. The demanding nature of the tech industry often leads to burnout and stress. It is important for employees to set boundaries, take breaks, and prioritize self-care. Regular exercise, healthy eating, and fostering hobbies outside of work can help individuals recharge and bring their best selves to their professional endeavors.

In conclusion, building healthy habits is essential for long-term success in the tech industry. By adopting a continuous learning mindset, effective time management, nurturing professional relationships, and maintaining a healthy work-life balance, employees can navigate their career paths with confidence and achieve professional growth in the digital world.

Finding Fulfillment and Purpose in Your Tech Career

In today's rapidly evolving digital world, the tech industry offers endless opportunities for career growth. However, professional success alone may not guarantee true fulfillment and purpose in your tech career. To truly excel and find meaning in your work, it is vital to uncover your passions and align them with your professional goals. This subchapter explores the importance of finding fulfillment and purpose in your tech career and provides practical strategies to help you navigate this journey.

One of the first steps towards finding fulfillment is understanding your personal motivations. Ask yourself what drives you to work in the tech industry. Is it the desire to solve complex problems, create innovative solutions, or make a positive impact on society? By identifying your core motivations, you can begin to align your career choices with your passions, leading to a more fulfilling professional journey.

Another crucial aspect of finding purpose in your tech career is setting meaningful goals. Instead of solely focusing on climbing the corporate ladder or chasing monetary rewards, consider setting goals aligned with your personal values and aspirations. This could involve mastering a new programming language, leading a team on a socially impactful project, or becoming a subject matter expert in a specific tech domain. Setting these goals not only provides a sense of purpose but also helps you stay motivated and engaged in your work.

Furthermore, finding fulfillment in your tech career involves continuously learning and growing. The digital world is ever-evolving, and keeping up with the latest trends and technologies is essential. Seek out opportunities for professional development, whether it be attending conferences, enrolling in online courses, or participating in

industry forums. By expanding your knowledge and skillset, you not only enhance your value as an employee but also open doors to new and exciting career prospects.

Lastly, don't underestimate the power of networking and building meaningful connections. Surrounding yourself with like-minded professionals who share similar values and ambitions can be highly motivating and provide invaluable support and guidance throughout your career journey. Engage in tech communities, join relevant professional groups, and actively seek mentorship opportunities to foster these connections.

In conclusion, while career growth is important, finding fulfillment and purpose in your tech career is equally crucial. By understanding your motivations, setting meaningful goals, continuously learning, and building a strong network, you can navigate the path to professional growth in the digital world while also experiencing a deep sense of fulfillment and purpose. Remember, the journey towards a fulfilling tech career is a lifelong one, so embrace the challenges, seize opportunities, and never stop striving for personal and professional excellence.

Conclusion: Mastering Your Tech Career Journey

Congratulations! You have reached the end of this book, "Tech Career Mastery: Navigating the Path to Professional Growth in the Digital World." Throughout this journey, we have explored various aspects of your tech career and provided you with valuable insights and strategies to help you achieve success. As you conclude this chapter, it is important to reflect on the key takeaways and prepare yourself to master your tech career journey.

In today's rapidly evolving digital world, a tech career offers immense opportunities for growth and innovation. However, to make the most of these possibilities, you need to be proactive, adaptable, and continuously learning. The path to career growth in the tech industry is not a straight line but a dynamic and ever-changing landscape. It requires constant effort to stay ahead of the curve and navigate the challenges that come your way.

One of the fundamental principles we discussed throughout this book is the importance of developing a growth mindset. Embracing a growth mindset means seeing every setback as an opportunity to learn and improve. It means being open to new ideas, seeking feedback, and continuously expanding your skillset. By adopting a growth mindset, you position yourself to take advantage of new technologies, trends, and opportunities that arise in your tech career.

Another crucial aspect we explored is the significance of building a strong professional network. Your network can provide you with valuable connections, mentorship, and opportunities that can accelerate your career growth. Actively engage in networking events, industry conferences, and online communities to connect with like-minded professionals and experts in your niche. Remember, your network is not just about what you can get from others, but also what you can contribute and give back to the community.

Continuous learning and upskilling are the keys to staying relevant in the tech industry. As technology continues to advance, it is crucial to invest in your professional development. Seek out opportunities for training, certifications, and workshops that can enhance your skillset. Stay updated with the latest industry trends and technologies, and be proactive in acquiring the knowledge and expertise required for your desired career path.

Lastly, never underestimate the power of perseverance and resilience. The tech industry is filled with challenges and obstacles, but those who are determined and resilient are the ones who ultimately succeed. Embrace failure as a stepping stone towards success and never give up on your dreams and aspirations.

In conclusion, mastering your tech career journey requires a combination of a growth mindset, a strong professional network, continuous learning, and resilience. By implementing the strategies and insights provided in this book, you are well-equipped to navigate the ever-changing digital landscape and achieve significant career growth. Embrace the challenges, seize the opportunities, and remember that your tech career is a lifelong journey of learning and development. Good luck!